ESPIONAGE BLACK BOOK FIVE

In this series of technical monographs:

Espionage Black Book One: Intelligence Databases Explained

Espionage Black Book Two: Codes and Ciphers Explained

Espionage Black Book Three: Surveillance Explained

Espionage Black Book Four: Open-Source Intelligence Explained

ESPIONAGE BLACK BOOK FIVE:
Basic Intelligence Explained

Henry W. Prunckun

Bibliologica Press

Espionage Black Book Five:
Basic Intelligence Explained

Copyright © 2022 by Henry W. Prunckun

This book is copyright.
Except in the case of study, research, or review,
no portion of this book may be reproduced by any means
without the prior written permission of the author.

ISBN 978-0-6452362-7-9

 A catalogue record for this book is available from the National Library of Australia

For information on all Bibliologica Press's publications,
visit our Web site at bibliologica.com

Bibliologica Press
P.O. Box 656
Unley, South Australia, 5061
Australia

CONTENTS

Basic Intelligence 1

Taxonomy of Intelligence 8

Collection and Archiving 14

Collection Development 26

Information Standards and Validation 38

About the Author 45

Index 46

— CHAPTER ONE —
BASIC INTELLIGENCE

The word *basic* describes something fundamental, or something that forms the base for further development or establishes the underlying foundation to what could later be enhanced. Take, for example, the military term *basic training*. It refers to the rudimentary subjects military recruits study when joining the armed services. Asked to describe this level of training, one might respond with the adjective "elementary."

When it comes to *basic intelligence*, one might, naturally, assume that this category of intelligence is rudimentary. But it is not. Basic intelligence is a distinct category of intelligence that requires the same level of planning and consideration as other categories of intelligence. This is because it is:

> ...factual information which results from the collection of encyclopedic information of more or less permanent or static nature and general interest which, as a result of evaluation and interpretation, is determined to be the best available.[1]

---

1. Chief of Naval Education and Training, prepared by Terry L. Schroeder, *Intelligence Specialist 3 & 2, Volume 1* (Washington, DC: U.S. Government Printing Office, 1983), p. AIII-15.

Basic intelligence refers to factual information[2] that is accurate at a particular point in time (or a range of dates) or is lasting. Basic intelligence is used for background briefings, evaluations, assessments, estimates, and counterintelligence investigations. Yet, as with many terms, basic intelligence is also used as an adjective to describe some entry-level intelligence courses[3] and lower-level analytical positions,[4] which may cause confusion unless the term is read in context with the topic being discussed.

The purpose of basic intelligence is to collect, storge, and index reference material about intelligence targets or subjects of interest. Targets might be people, companies, organizations, countries, geographic locations, or issues of concern. The idea behind amassing such an inventory of reference material is for analysts to retrieve it when a *consumer* needs advice or provide factual information directly to consumers.

The Office of the Secretary of Defense, Military Departments, Joint Staff, combatant commands, Defense agencies, and any other Department of Defense use the

---

2. That is, there are no opinions, inferences, conclusions, probabilities, estimates, or the like contained in these data.

3. These type of course usually teach newcomers basic analytical skills that allow them to produce reasoned descriptive intelligence reports and briefings.

4. For instance, the position description for an Intelligence Research Specialist, stated, in part, "Judgment is exercised in adapting *basic intelligence techniques* to particular situations and in evaluating validity and pertinence [i.e., reliability] of data and reports." (Emphasis added) U.S. Office of Personnel Management, *Position Classification Standard Flysheet for Intelligence Series, GS-0132* (Washington, DC: Office of Personnel Management, 1960), p. 14.

following definition as their term, except when communicating with their North Atlantic Treaty Organization (NATO) partners, at which time they use NATO terminology.[5]

> Fundamental intelligence concerning the general situation, resources, capabilities, and vulnerabilities of foreign countries or areas which may be used as reference material in the planning of operations at any level and in evaluating subsequent information relating to the same subject.[6]

NATO defines *basic intelligence* as "Intelligence derived from any source that may be used as reference material for planning and as a basis for processing subsequent information or intelligence."[7] NATO goes on to say that "Basic intelligence is fused from all available data, information, joint intelligence, surveillance and reconnaissance results, single-source intelligence and all-source intelligence and it is fundamental to current intelligence."[8]

It should be pointed out that the U.S. military also refers to basic intelligence as *general military intelligence*, which it abbreviates as GMI. The military's definition of GMI is narrower than what an intelligence agency would use; this is because its focus is on military aspects.

---

5. U.S. Department of Defense, *DoD Dictionary of Military and Associated Terms* (Washington, DC: Department of Defense, 2004), p. i. It is interesting to note that the 2021 edition of this publication does not contain an entry for *basic intelligence*.

6. U.S. Department of Defense, *DoD Dictionary of Military and Associated Terms*, p. 62.

7. NATO, *NATO Glossary of Terms and Definitions, AAP-06, Edition 2021* (Brussels, Belgium: NATO, 2021), p. 18.

8. NATO, *NATO Glossary of Terms and Definitions*, p. 18.

Nevertheless, the definition still speaks of factual information:

> Intelligence concerning the military capabilities. of foreign countries or organizations, or topics affecting potential United States or multinational military operations.[9]

Figure 1—View of NATO's library in Brussels, Belgium. Courtesy of NATO.

One of the Five Eyes[10] nations, the United Kingdom, defines basic intelligence as:

---

9. U.S. Department of Defense, *DoD Dictionary of Military and Associated Terms* (Washington, DC: Department of Defense, 2021), p. 91.

10. "The Five Eyes intelligence community grew out of twentieth-century British-American intelligence cooperation. While not monolithic; the group is more cohesive than generally known. Rather than being centrally choreographed, the Five Eyes group is more of a cooperative, complex network of linked autonomous intelligence agencies, interacting with an affinity strengthened by a profound sense of confidence in each

...intelligence on any subject that may be used as reference material for planning and as a basis for processing subsequent information or intelligence. Basic intelligence includes details of orders of battle, equipment capabilities, personalities, infrastructure, socio-political, economic, and environmental aspects. We derive basic intelligence through routine monitoring or on a contingency basis. Some UK intelligence agencies use the term 'building-block intelligence' when referring to basic intelligence.[11]

Here, we see a third term in use—*building block intelligence*. This term, which is also used by New Zealand intelligence services,[12] should not be confused with the children's toy by the same name[13] or the term

---

other and a degree of professional trust so strong as to be unique in the world." James Cox, *Canada and the Five Eyes Intelligence Community* (Calgary, Alberta: Canadian Defence and Foreign Affairs Institute, 2012), p. 2.

11. U.K. Ministry of Defence, *Understanding and Intelligence Support to Joint Operations, Third Edition, Joint Doctrine Publication 2-00* (Swindon, Wiltshire: Ministry of Defence, 2011), p. 2-9.

12. Inspector-General of Intelligence and Security, *Office of the Inspector-General of Intelligence and Security Inquiry into Possible New Zealand Intelligence and Security Agencies' Engagement with the CIA Detention and Interrogation Programme 2001–2009* (Wellington: Office of the Inspector-General of Intelligence and Security, 31 July 2019), p. 25.

13. "Zhejiang Jinma Arts & Crafts Co Ltd Files Chinese Patent Application for Bamboo Building Block Intelligence Toy," in *Global IP News* (New Deli: Pedia Content Solutions Pvt. Ltd., 2014).

building blocks of intelligence, which is used in cognitive psychology.[14]

In sum, basic intelligence refers to factual information, not "actionable intelligence."[15] This designation differentiates it from "raw intelligence," which "...refers to unevaluated intelligence information, generally from a single source, which has not been fully evaluated, integrated with other information, or interpreted and analyzed."[16]

Basic intelligence can be used as-is to provide customers with background information for, say, briefings, or it can be incorporated into secret studies where an intelligence report is produced. It can also help conduct counterintelligence investigations and "backstop"[17] counterespionage operatives' cover stories.[18] But before we examine basic intelligence in more depth, we need to understand how it differs from other categories of intelligence.

---

14. By way of example, Dimitri Van der Linden, et al., "Overlap Between the General Factor of Personality and Emotional Intelligence: A Meta-Analysis," in *Psychology Bulletin*, Volume 143, Number 1, 2017, pp. 36–52.

15. Also known as "finished intelligence."

16 Department of Justice, *FBI Information Sharing Report, 2010* (Washington, DC: FBI, Chief Information Sharing Officer, 2010), p. 21.

17. *Backstop* refers to the false background (i.e. a *legend*) that underpins an operative's cover. *Backstopping* is the verb. Bob Burton, *Top Secret: A Clandestine Operator's Glossary of Terms* (Boulder, CO: Paladin Press, 1986), pp. 12 & 66.

18. Jefferson Mack, *Running a Ring of Spies: Spycraft and Black Operations in the World of Espionage* (Boulder, CO: Paladin Press, 1996), p. 61.

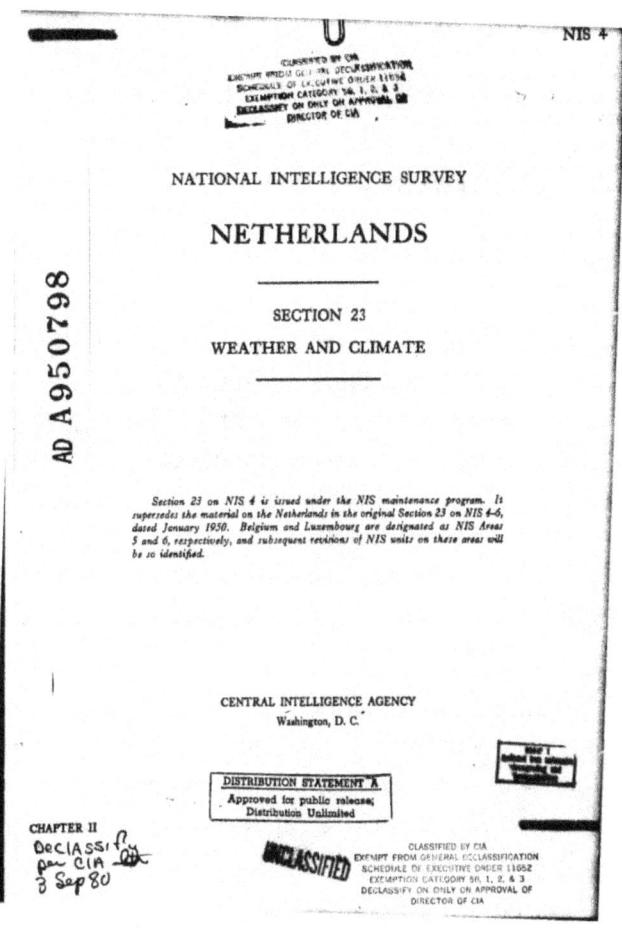

Figure 2—An example of a basic intelligence product is the CIA's National Intelligence Survey.[19] The National Intelligence Survey is an interagency report focusing on a specific country. Courtesy of the CIA.

---

19. "The basic unit of the NIS is the *General Survey*, which is now published [i.e., as of 1974] in a bound-by-chapter format so that topics of greater perishability can be updated on an individual basis." CIA, "General Survey: U.S.S.R.," in *National Intelligence Survey* (Langley, VA: CIA, 1974), preface.

— CHAPTER TWO —
TAXONOMY OF INTELLIGENCE

To some intelligence scholars, basic intelligence is viewed as either lacking importance—due to the allusion to being an endeavor that is "elementary"—or pretentious, because some interpret the term as insinuation[20] an "essential" nature to the practice.[21] It is neither; it is simply another category of intelligence practice. In this chapter, we will examine the taxonomy of intelligence to understand the relationship between each.

Intelligence can be categorized into four classes: 1) strategic; 2) warning; 3) basic; and the fourth comprises three subtypes—4a) tactical; 4b) operational; and 4c) current intelligence. These taxonomical categories are displayed graphically in Figure 3.

Strategic intelligence examines issues from the perspective of what might occur in the long term. What *long-term* means varies with agencies and even with projects within agencies. Six months might be considered long-term, or in other cases, a year or two might be the

---

20. i.e., synonymous.
21. Joseph W. Martin, "What Basic Intelligence Seeks to Do," in H. Bamford Westerfield, editor, *Inside the CIA's Private World: Declassified Articles from the Agency's Internal Journal, 1955–1992* (New Heaven, CT: Yale University Press, 1995), p. 207.

strategic study's horizon. Still, the same research process is involved, but with strategic intelligence, it may incorporate more complex methodologies and analytic techniques, as well as accessing information from a variety of sources.

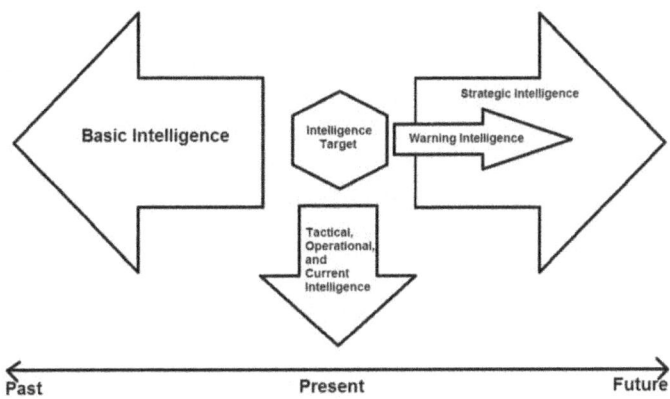

Figure 3—Graphical Representation of the Taxonomical Categories of Intelligence.[22]

The production of a strategic report could take weeks or months to complete, and then it is (or should be) subjected to some form of peer-view within the agency or by analysts in affiliated agencies. Review of the report's methods, the data's reliability, and the conclusions drawn ensures the validity of the research findings.

In short, strategic intelligence is forward-looking, and hence sometimes, it is described as being predictive. In this regard, within some circles, it is referred to as

---

22. Source: Hank Prunckun, *Methods of Inquiry for Intelligence Analysis, Third Edition* (Lanham, MD: Rowman & Littlefield, 2019), p. 17.

estimative intelligence.[23] Yet, any emphasis placed on prediction must be tempered: "Analysts are estimators [of the future], not clairvoyants. In the best of circumstances, a well-trained analyst cannot read an opponent's mind. Analysts can explain a trend or understand motive, but they will not know everything."[24] As such, consumers of strategic reports must be aware of these limitations and understand that at the heart of intelligence work is the *reduction of uncertainty*.[25]

Although operational intelligence is a forward-looking type of analysis, it is shorter-range or time-limited but is longer-term than tactical intelligence. Operational intelligence studies patterns or the operational mode activities of a target that provides a broader perspective than tactical intelligence.

Tactical intelligence provides immediate insights regarding a target or its activities to customers. These insights are aimed at providing decision-makers with more than factual information (i.e., basic intelligence, though it may include such data), as estimates of vulnerability, likelihood, consequences, impact, risk, outcome, intent, capability, and so forth.

Depending on the agency, tactical intelligence is known by other names, such as in law enforcement, where it is called *field intelligence* and *line intelligence*. In the military, it can be referred to as *combat intelligence*.

---

23. Hank Prunckun, *Methods of Inquiry for Intelligence Analysis, Third Edition*, p. 16.
24. Frank J. Cilluffo, Ronald A. Marks, and George C. Salmoiraghi, "The Use and Limits of U.S. Intelligence," in *Washington Quarterly*, 2002, Volume 25, Number 1, p. 72.
25. Hank Prunckun, *Methods of Inquiry for Intelligence Analysis, Third Edition*, pp. 4 & 9.

Regardless, all of these terms are accommodated under the umbrella name of *current intelligence*. Irrespective of variations, this discussion has opted to refer to these distinctions as *tactical intelligence*.

Warning intelligence notifies policymakers of developments of a sudden nature. In the realm of national security and military intelligence, events might be a military coup or an invasion by foreign troops, such as Russia's invasion of Ukraine in February 2022.

It is argued here that basic intelligence predates all of these intelligence categories. It is asserted that basic intelligence came of age in 1230 when a Parisian abbot and an Oxford polymath simultaneously invented a system for indexing information. The monk, Hugh of Saint-Cher, created a word index for the Bible with the help of his fellow monks. Some 10,000 unique words/phrases were located in the text and assigned page numbers.[26]

Robert Grosseteste, a thirteenth-century philosopher, theologian, and scientist, created a more modest index of some 400 words. He called a "grand table" that contained words of interest across all of his readings.[27]

Although Sun Tzu refers to the art of intelligence at different places in his *The Art of War*, a close reading of these passages, however, shows that he is describing the use of basic intelligence. Take, for instance, his advice,

---

26. John B. Friedman, *John de Foxton's Liber Cosmographiae (1408): An Edition and Codicological Study*, (Leiden, The Netherlands: E.J. Brill, 1988), p. xli.

27. Dennis Duncan, *Index, A History of*, (London: Allen Lane, 2022).

If you know the enemy and know yourself, you need not fear the result of a hundred battles. If you know yourself but not the enemy, for every victory gained you will also suffer a defeat. If you know neither the enemy nor yourself, you will succumb in every battle.[28]

Figure 4—The card catalog replaced lists and tables, but itself was replaced by digital indices. Courtesy of the author.

It is asserted that Sun Tzu is advocating for factual information about the enemy's position and capabilities to be able to infer their plans (i.e., the product of using these data in the production of tactical intelligence). Likewise, it is posited that he is advising that one needs to understand one's strengths and shortcomings to formulate both offensive and defensive strategies.

The Christian Bible also discusses the use of basic intelligence in decision making:

---

28. Sun Tzu, *The Art f War: Complete and Unabridged* (Mumbai: Wilco Publishing House, 2008), p. 23.

17. Moses sent them to spy out the land of Canaan. He said to them, 'Go up there into the Negev. Then go up into the hill country. 18. See what the land is like. See if the people who live in it are strong or weak, and if they are few or many. 19. Find out if the land they live in is good or bad. See if the cities they live in are open or if they have walls. 20. Find out if the land is rich or poor, and if there are trees in it or not. Then try to get some of the fruit of the land.' Now this was the gathering time of the first grown grapes.[29]

Parsing these data collection directives, we can see that Moses was looking for factual information, not interpretation of what these data meant (i.e., tactical. operational, or strategic intelligence). It is also interesting to note that amongst the information to be collected, he was interested in facts in the form of specimens, or as we will refer to them—artifacts—"…try to get some of the fruit of the land."[30]

It is therefore postulated that basic intelligence is the oldest type of intelligence. It is encyclopedic, comprehensive, and strives to be exhaustive. It is in contrast to the other three categories of intelligence that are, arguably, estimative in nature. Therefore, an encyclopedic holding of facts, objects,[31] images, and visual representations require consideration in the way these data are collected and stored so that they can be quickly retrieved for use in research projects.

---

29. *The Holy Bible*, Numbers 13: 17–20.

30. Factual information can take many forms, from information to physical objects, and visual representations (e.g., maps, photographs, drawings, artworks, graphical designs, etcetera.

31. An example of such an object is cited on page 13.

— CHAPTER THREE —
COLLECTION AND ARCHIVING

It would be hard to see a case where some pieces of basic intelligence information does not find its way into finished intelligence products. This proposition underscores its essential nature and needs a strategy to guide the collection and archiving of these data. The doctrine of basic intelligence comprises two principles: 1) the need to have clear aims, and 2) the methods of achieving these aims need to be efficient.

The aim of basic intelligence is to act as a "…base for intelligence products in support of planning, policy making, and military operations."[32]

> Basic intelligence consists primarily of the structured compilation of geographic, demographic, social, military, and political data on foreign countries. This material is presented in the form of maps, atlases, force summaries, handbooks, and, on occasion, sandtable models of terrain.[33] The Directorate of Intelligence in CIA, an NIMA [National Imagery and Mapping Agency], and the Directorate of Intelligence of

---

32. Director of Central Intelligence, *A Consumer's Guide to Intelligence*, p. 41.

33. A sandtable model is an example of where an object/artifact or specimen that may be included in a knowledge repository's collection of information as discussed on page 12. Specimens can also be scanned to produce a three-dimensional digital image that can be stored, retrieved, and viewed electronical.

intelligence production in DIA [Defense Intelligence Agency] are major producers of this kind of material.[34]

Arguably, the most widely known publicly available publication is *The CIA World Factbook*.[35] This and other basic intelligence publications provide:

> Factual, fundamental, and relatively permanent information about all aspects of a nation—physical, social, economic, political, biographical, and cultural…[36]

However, to fulfill this aim, an efficient system of data collection and archiving needs to be created. A *library* immediately presents itself as a mechanism for such a venture.

Depending on the type (i.e., class) of intelligence[37] and the specific agency, an archive may be organized administratively under different organizational names. For example, the U.S. Central Intelligence Agency once referred to its basic intelligence group as the Central

---

34. Director of Central Intelligence, *A Consumer's Guide to Intelligence*, p. 4.

35. Central Intelligence Agency, *The CIA World Fact Book* (New York: Skyhorse Publishing, 2019).

36. Director of Central Intelligence, *A Consumer's Guide to Intelligence*, p. 41.

37. That is to say, national security intelligence, military intelligence, law enforcement intelligence, business intelligence, private sector intelligence, and non-government organization (NGO) intelligence. Hank Prunckun, *Methods of Inquiry for Intelligence Analysis, Third Edition*, pp. 26–30, and Troy Whitford and Henry Prunckun, "Discreet, Not Covert: Reflections on Teaching Intelligence Analysis in a Non-Government Setting," *Salus Journal*, Vol. 5, No. 1, 2017, pp.48–61.

Reference Service.[38] After an organizational restructure, the CIA changed the name to the Office of Support Services, which housed the Directorate of Intelligence's "knowledge repositories."[39]

Although an administrative unit's name will change in response to a variety of managerial demands, the function of these units with regard to curating basic intelligence will remain the same; that is:

> ...informing policy makers of developments which might affect the national interest ... [and] provides what has been called 'a bridge between the past and the future.' The past is in embodied, essentially, in basic descriptions of the National Intelligence Surveys. The future is represented in that function of intelligence which puts the greatest demand upon the capabilities of intelligence analysts.[40]

In the early years of libraries, their function was to preserve the written word but evolved into a repository for readers and researchers.[41] Typically, libraries house books, academic journals, magazines, newspapers, trade periodicals, films, and other artifacts, like computer programs, etcetera, which can usually be borrowed and read by "members" of the library. A reference library that houses basic intelligence contains the same material

---

38. Victor Marchetti and John D. Marks, *The CIA and the Cult of Intelligence* (New York: Alfred A. Knopf, 1974), p. 68.

39. Director of Central Intelligence, *A Consumer's Guide to Intelligence* (Darby, PA: Diane Publishing Company, 1999), pp. 13–14.

40. Harry Howe Ransom, *The Intelligence Establishment* (London: Oxford University Press, 1970), p. 35.

41. Andrew Pettegree and Arthur der Weduwen, *The Library: A Fragile History* (London: Profile Books, 2021).

forms, but these materials generally stay in the section of the library dedicated to reference material.

The common method for assessing the "power" of a library is in the number of volumes[42] it holds. By the end of the Second World War, libraries were recognized as being important to national security.[43] So, it is no wonder that in March 2022, the CIA acknowledged that its library held about 125,000 books and subscribed to some 1,700 periodicals.[44] Its "Historical Intelligence Collection," according to the CIA, contained some 25,000 books (see Figure 5).

The two most used systems for organizing library resources are the Dewy Decimal Classification system[45] and the Library of Congress Classification system.[46] Both systems organize material according to the subject.

Under each system, reference material will be found, normally, in a dedicated section of the library. The types of material that comprise the bulk of this type of material include dictionaries, almanacs, yearbooks, bibliographies, autobiographies and memoirs, indices, encyclopedias, statistical reports, maps, atlases, manuals, policy documents, foreign and domestic telephone directories,

---

42. i.e., individual works in the collection.

43. Alex Boodrookas, "Total Literature, Total War: Foreign Aid, Area Studies, and the Weaponization of U.S. Research Libraries," in *Diplomatic History*, No. 43, 2019, pp. 332–352.

44 Central Intelligence Agency, "About the CIA Library" (https://www.cia.gov/legacy/headquarters/cia-library/), accessed March 17, 2022.

45. Shortened to the "Dewey Decimal System." The system's ten classifications encompass all categories of knowledge.

46. This system is used by research and academic libraries.

annual reports, *Who's Who*-type biographical sources, foreign diplomatic lists, and many more.[47]

Figure 5—View of some of the books that comprise the CIA's "Historical Intelligence Collection." Courtesy of the CIA.

Although this list appears extensive, especially if one considers that there can be sub-categories within each of these forms of reference material, such as language, country, and timeframes, it is indicative. Moreover, to remain useful, basic intelligence needs to be comprehensive and must also be organized for quick

---

47. Biographical publications are valuable when it comes to targeting individuals as was done in 2022 when the United States and its allies imposed sanctions on several Russian oligarchs who had close ties to Putin following Moscow's invasion of Ukraine. Biographical information is also maintained on opposition intelligence officers. Vladimir Kuzichkin, *Inside the KGB: Myth and Reality* (London: Andre Deutsch, 1990), p. 52.

access.[48] Above all, the collection must be continually updated to ensure new information is incorporated into the collection.

The subject material that finds its way into a basic intelligence library's holding is determined by users' needs (based on current intelligence needs and priorities). These needs are reflected by the environment in which an agency operates. Information classifications are structured by what is termed an intelligence *typology*, and there are six types of intelligence.

The first type is *national security* intelligence, which is sometimes referred to as *foreign policy intelligence*,[49] or simply *foreign intelligence*.[50] It is the most prominent collector of basic intelligence because its subjects are extensive. Some of the topics that immediately present for collection are political issues within a targeted foreign country, along with issues regarding its health, education, and welfare institutions, its social problems, and its legal environment.

Additional subjects that concern national security intelligence include "…food production and distribution, world resources (e.g., oil and potable water), international trade relationships, world migration patterns and changes in the ethnic composition of nations, as well as the state of the global monetary order. Without a doubt, [national security intelligence seeks to collect basic intelligence] …

---

48. Donald Cleveland and Ana Cleveland, *Introduction to Indexing and Abstracting, Fourth Revised Edition* (Santa Barbara, CA: Libraries Unlimited, 2013).

49. Ray S. Cline, "Policy Without Intelligence," in *Foreign Policy*, Winter, 1974–1975, Number 17, pp. 132 & 133.

50. Patrick F. Walsh, *Intelligence and Intelligence Analysis* (New York: Routledge, 2011), pp. 10 & 12.

on foreign technological developments, nuclear matters, and almost anything to do with foreign weapons production, defense industries, defense installations, and military capabilities."[51]

The second type is *military intelligence*. Although there may be some overlap with national security intelligence in the subject matter collected, this type of intelligence focuses on military matters. Examples of the reference material collected include historical data of a nation's military "...strength, capabilities, and vulnerabilities, as well as information on weather and terrain [in the region]."[52] These reference materials cover all nations, whether friendly, hostile, or neutral, because alliances change, and new coalitions are created as global political developments unfold. Therefore, it is important to have these details available for incorporating into intelligence assessments.

Law enforcement intelligence is the third type of intelligence. Note that this is broader than "police intelligence" because it is not only police forces that enforce laws. There are regulatory and compliance agencies with powers to conduct investigations and prosecutions. And, there is the often-overlooked sub-type of this classification; that is, "prison intelligence."[53]

Subjects vary according to the nature of the laws being enforced—from those on the international stage to those

---

51. Hank Prunckun, *Methods of Inquiry for Intelligence Analysis, Third Edition*, p. 27.

52. Joseph A. McChristian, *The Role of Military Intelligence: 1956–1967* (Washington, DC: Government Printing Office, 1974), p. 3,

53. Patrick F. Walsh, *Intelligence and Intelligence Analysis*, p. 36.

of concern to local communities, though there is overlap with targets like the six types of intelligence. For instance, take organized crime; it might be a target of a national security intelligence research project (e.g., where criminals are financing terrorism), a military intelligence interest (e.g., sales of arms on the black market); and with regard to law enforcement, political corruption, smuggling, illicit drugs, money laundering, kidnapping, tax evasion, and wire fraud.

Business intelligence is the fourth type of intelligence. Some practitioners of the craft use alternative names, such as *competitor intelligence, competition intelligence,* and *corporate intelligence.* In the past, the media has exposed some unethical methods of a few of these practitioners, which has detracted from the esteem that intelligence deserves. So, some businesses have tried to "soften" the image of spying by adopting names like *market research, product research,* or *customer/client research.*[54]

Regardless, business intelligence is concerned with the acquisition of trade-related data and commercial information held by competing firms. Therefore, the subject material that is collected relates to all aspects of competitors' processes, from the acquisition of raw materials to the distribution and sale of the finished products. The same applies to businesses that offer services.

Yet, it is not only businesses that engage in this form of data collection; foreign nations' national security and military agencies also target overseas businesses to

---

54. After all, *intelligence* is research where some part of the process is secret. See, Hank Prunckun, *Methods of Inquiry for Intelligence Analysis, Third Edition,* p. 4.

acquire confidential information. Take the case during the COVID-19 pandemic where it was alleged that two Chinese nationals—Li Xiaoyu and Dong Jiazhi—working for and assisted by the Chinese Ministry for State Security[55] penetrated the computer networks of bio-research laboratories to steal vaccine information.[56]

> ...they researched vulnerabilities in the networks of biotech and other firms publicly known for work on COVID-19 vaccines, treatments, and testing technology. Their victim companies were located all across the world, including among other places the United States, Australia, Belgium, Germany, Japan, Lithuania, the Netherlands, South Korea, Spain, Sweden, and the United Kingdom.[57]

The fifth type is *private-sector intelligence*, which spans several areas of intelligence work.[58] It is secret research conducted by entities that do not fall under the umbrella of government service. Practitioners often come to this

---

55. The Ministry for State Security is the People's Republic of China's foreign intelligence agency. David Wise, *Tiger Trap: America's Secret Spy War with China* (Boston: Houghton Mifflin Harcourt, 2011), p. 6.

56. Julian E. Barnes, "U.S. Accuses Hackers of Trying to Steal Coronavirus Vaccine Data for China," *The New York Times* (https://www.nytimes.com/2020/07/21/us/politics/china-hacking-coronavirus-vaccine.html), accessed March 14, 2022.

57. William D. Hyslop, James A. Goeke, and Scott K. McCulloch, *United States of America Vs Li Xiaoyu and Dong Jiazhi, Criminal Indictment*, Case 4:20-cr-06019-SMJ, ECF, No. 15, filed 7 July 2020 in the United States District Court for the Eastern District of Washington.

58. Hank Prunckun, *Methods of Inquiry for Intelligence Analysis, Third Edition*, p. 26.

sector after retiring from national security, military, or law enforcement intelligence positions.[59]

Whether they are firms or individual contractors, these operatives are essentially freelance spies. This description should not be read as derogatory; it is merely a matter-of-fact phrase to explain the basis for how the sector operates. A private intelligence agent may be hired by a business or a private citizen, or even a government or military unit. Non-government organizations may also employ them.

Although private-sector intelligence operatives can move from one specialty to another, these private agents will likely focus their expertise on a narrow field of inquiry, so the subject material they collect will be concentrated on that area of study.

In practice, private sector agents are likely to either rely on basic intelligence provided by their government employer or access these historical data items via online searches because there are a sizeable number of open and semi-open sources of information. By way of example, let us look at the "Russell J. Bowen Collection on Intelligence, Security and Covert Activities" held by the Georgetown University Library, Washington, DC. It is a:

> ...collection of books on the subjects of intelligence, spying, covert activities, and related fields. The collection of more than 17,500 titles includes works on cryptography, signal intelligence, tradecraft of all kinds, and the application of modern technology to intelligence gathering. The separately maintained Bowen Spy

---

59. Anthony D. Manley, *The Elements of Private Investigation: An Introduction to the Law, Techniques, and Procedures* (Boca Raton, FL: CRC Press, 2010).

Fiction Collection contains more than 3,500 titles in the spy fiction genre.[60, 61]

This does not suggest that private operatives do not collect their own basic intelligence, but operating as an outworker, makes it is less likely.[62] As Leonard M. Fuld, a pioneer in business intelligence, advocated, a basic intelligence library for a private-sector operator should be "lean and mean."[63]

Fuld suggests only a handful of reference texts are needed, and in lieu of more comprehensive collection, business intelligence agents access the material they need via the library system, specifically, university libraries.[64]

The last type of intelligence is *non-government intelligence*. It also goes by the name *humanitarian intelligence*, which assesses risks (e.g., to the physical and psychological well-being of employees and clients), and

---

60. Georgetown University Library, "Russell J. Bowen Fund" (https://library.georgetown.edu/giving/endowments/bowen), accessed March 18, 2022.

61. Given the size of the Bowen Collection in 2022, it is comparable to the same size of the CIA's "Historical Intelligence Collection."

62. Henry W. Prunckun, "Digest Notes: An Interview with George Martin Barringer, Special Collections Librarian, Russell J. Bowen Collection of Works," in *Law Enforcement Intelligence Digest*, Volume 6, Number 1, 1991, pp. 48–50.

63. Leonard M. Fuld, *Competitor Intelligence: How to Get It— How to Use It* (New York: John Wiley, 1985), p. 29.

64. Leonard M. Fuld, *Competitor Intelligence: How to Get It— How to Use It*, p. 27.

valuates political,[65] military, social, and cultural trends in the area of operation.[66]

As with their colleagues in private-sector intelligence, these agents are also likely to rely on basic intelligence provided by their coordinating body/funding body (e.g., United Nations), or access these historical data items via any number of open and semi-open online databases. Again, this does not suggest that these operatives do not collect their own basic intelligence, but being freelancers, it is less likely.[67]

---

65. Here, we use the term *political* to also refer to the situation's religious milieu.

66. Andrej Zwitter, *Humanitarian Intelligence: A Practitioner's Guide to Crisis Analysis and Project Management* (Lanham, MD: Rowman & Littlefield, 2016), p. 25.

67. Leonard M. Fuld, *Competitor Intelligence: How to Get It— How to Use It*, pp. 27–29; and Anthony D. Manley, *The Elements of Private Investigation: An Introduction to the Law, Techniques, and Procedures*, p. 133.

— CHAPTER FOUR —
COLLECTION DEVELOPMENT

Information can be collected from a variety of sources. Many of these sources can be the same for the six types of intelligence practitioners we discussed in the previous chapter—national security, military, law enforcement, business, private sector, and non-government intelligence.

As such, it is conceivable that a single piece of information could have application in each of the functional intelligence environments. For example, information relating to a terrorist cell could be of interest to law enforcement intelligence officers who have a responsibility to prevent and deter such attacks on the homeland. Likewise, the same information could be of interest to:

- National security and military intelligence if the cell is based overseas or operates internationally;
- Business intelligence if the target of the terrorists is their industry or their facilities; or
- Groups employing private-sector intelligence, such as the anti-nuclear lobby, highlight the vulnerable nature of a nuclear facility about terrorism.

To highlight the various information sources and the diversity available, consider these indicative lists.

National Security Intelligence
- Open-source information (particularly the Internet);
- Liaison with friendly intelligence services;
- Clandestine operatives (official cover);

- Covert operatives (non-official cover);
- Recruited agents;
- Diplomatic missions and embassies;
- Surveillance aircraft;
- Surveillance satellites;
- Electronic intercepts (e.g., via listening posts);
- Defectors;
- University and independent research bodies; and
- Other government departments.

Military Intelligence

- Open-source information (particularly the Internet);
- Liaison with friendly intelligence services;
- Surveillance planes;
- Surveillance satellites;
- Electronic intercepts;
- Reconnaissance teams;
- Field operatives;
- Diplomatic missions and embassies;
- Defectors;
- Prisoners;
- Civilian inhabitants;
- University and independent research bodies; and
- Other government departments.

Law Enforcement Intelligence

- The public;
- Crime investigators (i.e., detectives and crime scene examiners);
- Patrol officers;
- Police records;
- The media;
- Businesses;
- Open-source information (particularly the Internet);
- Government departments and agencies;

- Overseas law enforcement agencies;
- Informants;
- Citizens;
- Covert surveillance (physical and electronic);
- Undercover operatives; and
- Other law enforcement agencies and government departments.

Business Intelligence

- Open-source information (particularly the Internet);
- A business's internal records;
- Information supplied by other businesses;
- The media, trade, and other open-source publications;
- Sales personnel;
- Customers;
- Distributors;
- Raw material and component suppliers;
- Government departments and agencies;
- A business's research and development section(s);
- University and independent research bodies;
- Market research surveys;
- Reverse engineering; and
- Covert physical surveillance (e.g., a hired private investigator).

Private Sector and Non-Government Intelligence

- An organization's internal records;
- Open-source information (particularly the Internet);
- Information supplied by other organizations;
- Liaison with friendly intelligence services;
- The media, trade, and other open-source publications;
- Staff;
- The public;
- Government departments and agencies;
- An organization's research section;

- University and independent research bodies;
- Surveys; and
- Covert physical surveillance (e.g., a hired private investigator).

Whether the data collected is large are small, the cache needs to be managed. This means there needs to be a library; in particular, a reference library that needs to be systematically organized.

This being the case, the information librarians collect is, *ipso facto*, determined by the users—the agency's analysts and consumers.[68] Although almost every form of information might be used in producing intelligence, the reports and briefings determine the users' needs—i.e., the inputs.

While a basic intelligence library is a specific type of collection, it is, after all, a library and is guided by the policies and practices adhered to by libraries in general. "All libraries are the product of a process of judicious selection."[69] Paramount is a collection development policy (CDP).

A collection development policy summarizes the principles that guide the library that has responsibility for the basic intelligence collection. Particular to our discussion is the criteria for selection. A generalized list

---

68. Recall, unlike finished intelligence products (e.g., operational assessments, national estimates, and other report types), basic intelligence products can be distributed to customers as-is. We earlier discussed the public release of *The CIA World Factbook* (New York: Skyhorse Publishing, 2019) as an example of a basic intelligence product.

69 Andrew Pettegree and Arthur der Weduwen, *The Library: A Fragile History*, p. 381.

of selection standards might include the following criteria, although individual agencies will no doubt tailor such a list to suit its mission:

- Relevance to the users' and customers' needs,[70] which are determined by the strategic goals of the agency;
- The authority (i.e., reliability and validity) of the material;
- Its enduring value to the collection; and
- Being mindful that taxpayers or shareholders fund the collection, value for money.

Preferences might be given to digital copies or to hardbound in preference to paperbound copies (because of the former's durability). The policy may also specify, in the case of open-source information, preferred suppliers, how donations to the library are processed, what to do with multiple copies (e.g., digitize them or place them in an archive), and how often will the collection be reviewed for lost or misplaced items.[71]

At this juncture, it would be valuable to look at what reference material constitutes a basic intelligence library. Although it would be unrealistic to produce a complete library catalog, a sketch might help transform the abstract ideas we have been examining into a more concrete form.

---

70. Needs can be assessed through users' suggestions, requests, and periodic needs analyses (e.g., via surveys), as well as through liaison with university libraries and research centers.

71. Australian Government Libraries Information Network, *Library Collection Development Policy* (Deakin, ACT: Australian Library and Information Association, 2021).

In outlining such a collection, it is important to point out that even though we discuss reference material in terms of physical books and journals, these materials exist in digital forms, searchable via a software interface. The terms also relate to digital databases and legacy forms of storage, such as microfilm and microfiche that have not yet been computerized.

Figure 6—Legacy forms of data storage, such as microfilm collections, will remain a feature in reference libraries until these records are digitized. Courtesy of the author.

The first group is dictionaries. These include both language and topical dictionaries. Language dictionaries include all the world's major spoken languages—English, French, Italian, German, Spanish, Portuguese, Russian, Hebrew, Greek, Persian, Arabic, Cantonese, Mandarin, and so on. It would also include a selection of colloquial and dialect dictionaries and an assortment of thesauri.[72]

---

72. These compilations of synonyms and related concepts extend beyond the complacent high school or college thesauri to many subjects—from anthropology to zoology.

Topical dictionaries would be selected for the subjects that are the target of research but would also incorporate allied and adjacent subjects of inquiries, such as law and political dictionaries; medical, nursing, and health dictionaries; psychology and psychiatric dictionaries; dictionaries of military terms and abbreviations; lexicons dealing with biology, chemistry, physics, mathematics, statistics, computing, technology, history,[73] government, criminology, penology, sociology, anthropology, theology, education, geography, mythology, and so forth.

As useful as online wikis are, encyclopedias are still the first choice for factual papers on a span of subjects. This is because web-based references are developed collaboratively by users. The fact that material can be created, edited, or deleted by a "community" of users is a double-edged sword. On the one hand, it makes for a free and accessible database of information easily accessed by anyone. Still, on the other hand, it lacks the quality control that academic peer-review brings to a commercially published encyclopedia.

Joseph W. Martin of the CIA, in his now declassified paper on basic intelligence, advocated that there are three criteria for considering a reference work as "excellent."[74] He proposed that excellence was determined by its

---

73. And, within each of these subjects, there are usually many subcategories. Take as an example, history. There may be topical dictionaries for ancient history, modern history, post-colonial history, European history, art history, military history, history of computing, etcetera.

74. Joseph W. Martin, "What Basic Intelligence Seeks to Do," in H. Bradford Westerfield, editor, *Inside CIA's Private World: Declassified Articles from the Agency's Internal Journal, 1955–1992* (New Haven, CT: Yale University Press, 1995), p.

"...systematic organization, clear and precise in its detailed presentation, [and] realistic in what it seeks to include."

With that caveat in place, a basic intelligence library would have various encyclopedias selected to serve analysts' needs. These would range from general encyclopedias, such as *Britannica*, that provide synopses on a range of topics to those that are subject-specific; for instance, encyclopedias on subjects previously mentioned about topical dictionaries.

Rated high in a basic intelligence collection would be research handbooks and guides compiled by the collection's librarians. These documents include subject information, instructions on finding practical study resources, as well as research tips (e.g., searching strategies and locations of material).

Bibliographies are often thought of as a list of publications that acknowledge the sources an author of a research paper has used. However, it also refers to book-length publications that subject-matter experts or research librarians compile. Other bibliographies may address single subjects, and some bibliographies feature annotations of the entries.

By way of example, take the distinguished *Scholar's Guide to Intelligence Literature: Bibliography of the Russell J. Bowen Collection*[75] that catalogs the holdings of this special collection on intelligence, security, and covert

---

75. Marjorie W. Cline, Carla E. Christiansen, and Judith M. Fontaine, *Scholar's Guide to Intelligence Literature: Bibliography of the Russell J. Bowen Collection* (Frederick, MD: University Publications of America, 1983).

activities held by the Georgetown University Library, Washington, D.C.

What is the purpose of including bibliographies in a basic intelligence library collection, especially if an intelligence analyst, targeting officer, staff operations officer, or other types of users, can conduct their own bibliographical search? The answer is that these reference books not only save searching time, which can be considerable, but they also separate the best sources from what can be a large volume of literary dross.

Because experts in the field compile subject-specific bibliographies, scholars have done the intellectual sorting and arranging of publications into sub-topics and related issues of interest. This intellectual oversight also applies to chronologies. Chronologies of events have been compiled by scholars on a long list of topics—terrorism being, perhaps, one of the most prevalent in recent decades.

Because basic intelligence comprises historical material,[76] there is a place for catalogs, directories, atlases, and maps. Take catalogs and directories; these are often issued regularly, and as such, they can be collected in date order—monthly, quarterly, yearly. In this way, users can consult changes over time, introduce new items, or remove others.

Catalogs are inventories of items systematically arranged, with descriptions and drawings or photographs of the items. The type of intelligence being practiced—national security, military, law enforcement, business,

---

76. Yes, basic intelligence is historical in nature, but it is not stagnant—it needs to be updated to account for changes, which helps ensure that it is as comprehensive as possible.

private sector, or non-government—will suggest the types of catalogs collected.[77]

Law enforcement intelligence might gather catalogs by gun and ammunition manufacturers, prescription drugs, chemicals, laboratory glassware, motor vehicles, and so on. As with the list of topical dictionaries we discussed earlier, catalogs can be just as diverse, notably when it comes to catalogs of statistical tables that many government bureaus produce.

Figure 7—*Who's Who* and similar publications are considered in the category of biographies. Courtesy of the author.

Diversity likewise applies to directories. These compilations organize information into, say, alphabetical order, such as the telephone directory. Although the phonebook is the most recognized directory, there are many others—ships, aircraft, manufacturers, suppliers of all descriptions, occupations, professions, and trades.[78]

---

77. And, not to forget, the needs of the users remain the chief reason for what is collected.
78. Directories also by other names, such as *registers*.

Many more could be listed, but the idea is obvious—there are directories of all descriptions. Not only that, like catalogs, these publications can be produced over time (e.g., year-on-year), by region, or by country.

We will take a minor detour in our conversation and look at biographies and yearbooks,[79] but note that other types of publications may present themselves in the same category.

These resources are of interest to users in a "Central Cover Division"[80] who, for instance, develop new identities for operatives deploying under non-official cover.[81] Or they may be of value to counterintelligence investigators who are checking the background and bona fides of a potential agent, a future employee, or a possible agency associate.

In this regard, family history material, such as family trees and privately published genealogical books, are beneficial in pursuing the fake identities of hostile agents and, conversely, help friendly operatives establish a more

---

79. For example, biographies, autobiographies, who's who-type publications, high school, college, and university yearbooks hold a wealth of information, including photographs. These publications lend themselves to digitization, adding users' ability to search images as well as data.

80. The term *Central Cover Division* is taken from the CIA-turned Cuban intelligence defector Philip Agee's book, *Inside the Company: CIA Diary* (New York: Stonehill Publishing Co., 1975), p. 21. It is used here as a generic label for any unit that specializes in this type of operational support.

81. Amaryllis Fox, *Life Undercover: Coming of Age in the CIA* (New York: Alfred A. Knopf, 2019), p. 118.

robust background for their cover, known as "backstopping."[82]

The final item we will discuss, although it is not the end of what could be found in a basic intelligence library, is maps and atlases. A *map* is a two-dimensional representation of an area. An *atlas* is several maps bound together. Areas can be as specific as a town map or as large as a map that depicts a country or continent. Areas change over time, so maps change too. Political borders change, the landscape changes, as do roads and utilities that service the area. Therefore, like catalogs and directories, maps and atlases are updated, and this process adds charts to the library's collection requirements.

Figure 8—Genealogical collections are helpful to investigators checking the identities of human assets and hostile agents. Courtesy of the author.

---

82. Almanacs are also useful because they provide information about the rising and setting times of the sun, the phases of the moon, as well as tables that show tidal, and other ancillary data.

— CHAPTER FIVE —
INFORMATION STANDARDS AND VALIDATION

The world is awash with information. Searching the Internet for information on any subject will return many sources; some report true and accurate data, but a percentage will be contradictory, wrong, misleading, or fabricated. The same applies to information in print. With the popularity of online self-publishing, this medium is subject to the same caveat.[83] Therefore, intelligence collectors need to have a standard against which data are assessed.

For such a standard, intelligence analysts look to the world of academia for guidance. In essence, this standard is found in the concepts of *reliability* and *validity*.

Reliability refers to the consistency of information. That is to say, if these data were to be collected by different methods (e.g., technical means or through human sources), would the information gathered be consistent? If the answer is "yes," it can be considered reliable.

Validity means that the information is accurate. If data are deemed to be of high validity, it means when it is used in a secret research project, this aspect of the study is likely to be valid. There are indicators of when data are

---

83. Academic journals and commercial publishing houses have manuscripts assessed by scholars in a process known as *peer-review*. Although not without shortcomings, the peer-review process helps ensure the reliability and validity of the research.

valid, and reliability is one indicator—if a data item is reliable, it is *probably* valid.[84]

Nevertheless, high reliability is not the only way to determine this, but other methods are not well defined. Still, there is a research axiom that states that information must reflect what it claims to be. As important as reliability is, validity is more important. This ensures an intelligence report's findings, discussion, and conclusions are also valid.

Before we discuss the types of reliability and validity, let us examine the criteria for evaluating information for inclusion into a basic intelligence library. These criteria should be seen as the first phase, or triage stage, in the overall process. This process applies to any source of information, whether it is a book, journal or magazine, website, or social media.

What is the source of information, and who was responsible for authoring it? What scholarly authority does the source's author hold? Indicators can include academic qualifications, the number of peer-reviewed papers published, and bibliometrics such as the number of citations for the author's publications.

If an organization of a foreign government collected the data (e.g., a bureau of statistics), questions would arise as to whether these entities are recognized as reliable sources with a reputation for quality control.

What was the purpose of publishing the data? A way of looking at this question is to assess whether the data is merely presenting facts, as in an annual report, or trying to

---

84. Zina O'Leary, *The Essential Guide to Doing Your Research Project, Third Edition* (Los Angeles: Sage, 2017), pp. 63–64.

be persuasive. To gauge the audience, look for clues: is the author/source speaking to fellow scholars or learned people, or to "anyone who will listen"?

Following close behind the question as to what the purpose of publishing the information is the question of objectivity. Does the source represent a lobby group, a special interest organization, or a political party, etcetera? One way to do this is to gauge whether certain data items are over or under-reported, exaggerated, or deemphasized.

Completeness and relevance are indicators when testing the information. Librarians and archivists should ask whether the data includes all it purports to represent or is an edited selection material from a fuller report (i.e., censored)? In addition, they should examine whether the data will address intelligence researchers' questions (or the background/context to these questions).

When it comes to collecting information from secret sources, a variation of this academic standard is applied. This standard is based on what has been termed the *Admiralty Grading System*.

This evaluation process also assesses the source's reliability and the information's validity, referred to under this system as *accuracy*. This system is known by various names by the American, Australian, British, Canadians and New Zealand militaries,[85] including "source and reliability matrix,"[86] "NATO System," and the "Admiralty

---

85. Chiefs of Staff, *Understanding and Intelligence Support to Joint Operations (JDP 2-00), Third Edition* (London: Ministry of Defence, 2011), pp. 3–21.

86. U.S. Department of the Amy, *Human Intelligence Collection Operations, Field Manual 2-22.3 (FM 34-52)* (Washington, DC: U.S. Army, 2006), pp. B1–B2.

Code." Regardless of the naming convention used, all are based on the so-called *Admiralty Grading System* developed by the British during World War Two.

A validation system grounded on the Admiralty Code's approach to data validity, but enhanced by this author, is shown in Tables 1 and 2. It is important to point out that in Table 2, there is a difference between *misinformation*, which is unintentional,[87] and *disinformation*, which is outright deception (i.e., intentional).

To explain the difference between the traditional Admiralty Code and what is being advanced in Table 3, we need to revisit the history of the scale. It was developed for tactical military purposes during the Second World War. Its purpose was to assign some level of certainty to information that would be the basis for combat operations (i.e., immediately or within a short time).

In this context, the Admiralty Grading System is fine. However, once an analyst starts to consider issues beyond operational matters, such as warning intelligence and strategic intelligence issues, then this scale presents inadequacies. This is because the scale is unable to consider unintentional misleading (misinformation) and intentionally deceptive information (disinformation) that are hallmarks of an opposition's counterintelligence operations. The "enhanced" scale presented here incorporates these two additional classes of information sources.

---

87. The term *unintentional* also incorporates information published by misguided authors, such as those who employ pseudoscience, poor logic, or adhere to conspiracy theories, divine prophesies, emotional arguments, or suffer from organic or substance-induced psychosis, etcetera.

Where data are collected for, say, a tactical research project, a validation process is performed on each piece of information entered into the agency's database. However, this process may be automated with basic intelligence, giving a published series of information (e.g., statistical tables released by a government's population bureau) a generic rating.

The process calls for each piece of data to be assigned an alphanumeric rating indicating the agency's degree of confidence in that piece of information.

Table 1—Source Reliability Codes.

	Admiralty Ratings	
Code	*Descriptors*	*Estimated Truth Based on Past Reporting*
A	Completely Reliable	100%
B	Usually Reliable	80%
C	Fairly Reliable	60%
D	Not Usually Reliable	40%
E	Unreliable	20%
F	Cannot be Judged	50%
G	Unintentionally Misleading	0%
H	Deliberately Deception	0%

As an example, imagine a field operative obtains a piece of information from an agent in place, but this agent is a new source that has never been exploited before. The reliability for this piece of information would therefore be F—the reliability cannot be judged. If the information obtained came from a source of previously provided information (for instance, another agent) and has proven to be truthful in almost every instance, then an accuracy code of 2 would be assigned. The combined code would be printed on the document to show its overall rating (in an electronic database, as digital marking would appear when viewing the item).

Customarily, an accuracy rating precedes the reliability code—for instance, F-2. Having said that, it needs to be pointed out that the ratings must be logical; assigning a rating of, say, E-2 (unreliable source but probably true) or H-3 (deliberately deceptive but possibly true) would raise questions in an analyst's mind about whether the evaluation process was rational.

Table 2.   Information Accuracy Estimates.

Admiralty Ratings		
Code	Descriptors	Estimated Probability of Truth
1	Confirmed	100%
2	Probably True	80%
3	Possibility True	60%
4	Doubtful	40%
5	Improbable	20%
6	Cannot be Judged	50%
7	Misinformation	0%
8	Disinformation	0%

Regarding the accuracy code 7 of Table 2—misinformation—the analyst should be cognizant they may obtain data that are *unintentionally* incorrect, illogical, or contradicted by other sources. In these cases, a code of 6 is appropriate. As for disinformation (code 8), these are data that are shown by other sources to be *deliberately* false or misleading (i.e., provided for the purposes of deception, perhaps as part of an opposition's counterintelligence operation).

Although the Admiralty Code is an attempt to establish an objective position for information, the product is the result of a subjective process. This is because human judgment plays a pivotal role.[88] When assigning a rating, the agency's librarians and archivists must consider such things as the accuracy of previous information provided by the source and the source's field capabilities (i.e., does the source have access and the ability to obtain what has been delivered?).

Evaluation is a difficult process but an important one because, without it, the lack of reliability or validity can adversely affect the findings of intelligence research projects.

---

88. Henry Prunckun, *Methods of Inquiry for Intelligence Analysis, Third Edition*, p. 46.

ABOUT THE AUTHOR

Dr Henry (Hank) Prunckun, BSc, MSocSc, MPhil, PhD, is an Adjunct Associate Research Professor at the Australian Graduate School of Policing and Security, Charles Sturt University, Sydney. He is a former Australian government intelligence analyst who spent much of his twenty-eight-year operational career in tactical intelligence and strategic research but also served operationally in security, investigation, and counterterrorism. During this time, he was conferred with two literary awards and a professional service award by the International Association of Law Enforcement Intelligence Analysts. After retiring from government service, he initially worked as a freelance private investigator, then spent over a decade as a research criminologist at Charles Sturt University examining transnational crime—espionage, terrorism, drugs and arms trafficking, and cyber-crime.

INDEX

Admiralty Grading System, 40, 41
Admiralty Ratings, 42, 43
Australia, 22, 40, 45
basic intelligence, 1–7; collection development, 26–37; defined, 2–6; validation, 38–44
biographical sources, 15, 18, 18n47
Bowen, Russell J., 23, 23–24, 24n60, 24n61, 24n62, 33, 33n75
building block intelligence, 5
business intelligence, 21, 21–22
Canada, 4–5n10
China. *See* Ministry for State Security
chronologies, 34
COVID-19, 22, 22n57
disinformation, 41, 43
Five Eyes, 4, 4n10
Georgetown University Library, 23, 24n60, 33–34
intelligence community, 4–5n10
intelligence; business, 21, 21–22; current, 11; ; estimative, 9–10; field, 10; humanitarian, 24–25, 25n66; law enforcement, 34, 20–21, 23, 26, 27–28, 34–35; line, 10; military intelligence, 3–5, 11, 15n37, 20–22, 20n52, 26, 27; non-government, 15n37, 23, 24, 26, 28–29, 35, ; operational, 8, 10,13, 29n68; prison, 20; private-sector, 22–24, 26; strategic, 8–10, 13, 30, 41; tactical, 8, 10–11, 12, 13, 41, 41; taxonomical categories, 8–9
Library of Congress, 17
Ministry for State Security, Chinese, 22, 22n55
misinformation, 41, 43
Moscow, 18n47
NATO System, 40
NATO, 3, 3n7, 3n8, 4, 40
New Zealand, 5, 5n12, 40
peer-review, 9, 32, 38n83, 39
Putin, 18n47
reliability, 2n4, 9, 30, 38, 38n83, 39, 40, 42, 43, 44
Russia, 11, 31
Sun Tzu, 11–12, 12n28
typology, 19
Ukraine, 11, 18n47
validity, 9, 30, 38–39, 38n83, 40–41, 44
Walsh, Patrick F., 19n50, 20n53

www.ingramcontent.com/pod-product-compliance
Lightning Source LLC
Chambersburg PA
CBHW022125040426
42450CB00006B/861